HOLDING THE DAM

Viruses of all kinds the world over, including AIDS, are mutating, growing more resistant to the drugs meant to combat them, threatening a new age of plagues. New research shows that malnutrition—in particular, deficiencies of selenium and vitamin E—promote such mutation. Dr. Richard Passwater explains this scientific breakthrough and explores the astonishing progress medicine is making in countering the ravages of AIDS and cancer.

Richard A. Passwater, Ph.D. is one of the most called-upon authorities in preventive health care. A noted biochemist, he is credited with popularizing the term "supernutrition" in such books as *Supernutrition: Megavitamin Revolution* and *The New Supernutrition*. His many other works include *Cancer Prevention and Nutritional Therapies* and *Pycnogenol*. Dr. Passwater lives in Berlin, Maryland, where he is the director of a research laboratory.

Dr. Passwater's research with selenium was first reported to the health food industry in the December 1971 issue of *Prevention* magazine. His research with selenium and other antioxidant nutrients had led to patents and a FDA IND application in 1970 for reducing free radical pathology and slowing the aging process. During 1970 through 1972, he found that selenium was a key antioxidant nutrient in reducing cancer incidence. In 1972, he filed for a patent involving selenium and other antioxidant nutrients in preventing cancer. In 1973, he published his research on selenium and other antioxidants in reducing cancer incidence in *American Laboratory*. He reviewed his selenium research and cancer in *Supernutrition: Megavitamin Revolution* in 1975 and *Cancer and Its Nutritional Therapies* in 1978. He published *Selenium as Food and Medicine* in 1980 and *Selenium Update* in 1987.

Selenium Against Cancer and Aids

The trace mineral that can make a life-or-death difference... and that many of us lack

Richard A. Passwater, Ph.D.

Keats Publishing, Inc.  New Canaan, Connecticut

DEDICATION

This book is dedicated to Herb Boynton, founder of Nutrition 21 of San Diego, California, for making selenium supplements available for public consumption.

Selenium Against Cancer and AIDS is intended solely for informational and educational purposes, and not as medical advice. Please consult a medical or health professional if you have questions about your health.

SELENIUM AGAINST CANCER AND AIDS

ISBN: 0-87983-784-5

Printed in the United States of America

Keats Good Health Guides are published by
Keats Publishing, Inc.
27 Pine Street (Box 876)
New Canaan, Connecticut 06840-0876

Contents

INTRODUCTION

In April 1996, after decades of supporting research, the relatively unknown trace mineral selenium was finally recognized by the scientific community as being the nutrient key to preventing cancer and halting the progression of HIV infection developing into AIDS. In six dynamic days, reports were presented confirming earlier evidence that selenium cuts the total cancer death rate in half, boosted the immune response in HIV-infected persons, and influenced the virulence of viruses and infections.

On April 16 at the annual meeting of the Federation of American Societies for Experimental Biology held in Washington, D.C., Dr. Larry Clark and his colleagues reported the results of a ten-year clinical study conducted in seven U.S. clinical centers showing that selenium significantly cut cancer incidence and cancer death rates.

Just the day before at this FASEB meeting, Drs. Melinda Beck and Orville Levander reported that viruses are more prone to mutate while infecting selenium-deficient or vitamin E-deficient persons (*FASEB* J. 10(3) A191 Abst. #1101, March 8, 1996). They reported that Coxsackie viruses that are normally harmless become mutated in selenium-deficient or vitamin E-deficient mice and then spread even in adequately nourished mice, to infect their hearts causing extensive damage.

This finding helps explain why strange diseases often arise in areas of human malnutrition and then spread to infect people all over the world. The selenium-deficient person, and perhaps people deficient in other antioxidant nutrients including vitamin E, in effect may change ordinarily harmless viruses into extremely deadly viruses. Antioxidants will be explained later, but it will suffice here to point out that antioxidants help protect both cells and viruses against mutation. However, viruses mutate a million times more often than body cells. A person deficient in either vitamin E or selenium facilitates a mutation in the viral genome and becomes a breeding ground for viral mutants that can then cause a worse, more virulent disease even in adequately nourished individuals.

Dr. Levander points out that this makes a strong argument that "we are our brother's keeper because we are not protected from what might be happening to malnourished people in Africa." Thus, if we allow our fellow man to be malnourished, we, in turn, may fall to new diseases spawned within their bodies and rapidly transmitted as new epidemics.

The selenium-deficient soils of large regions of Asia and Africa result in large populations of selenium-deficient persons. This may explain why so many severe viral epidemics arise at regular intervals in China. Some are even given names such as the Hong-Kong flu. HIV and Ebola are thought to have arisen in or near Zaire. Keshan Disease, a heart disease that affects the very young in China, is associated with selenium deficiency, but it may also have a possible association with a virus mutated as a result of selenium deficiency.

What do viruses such as HIV, Ebola and influenza have in common that make them prone to such drastic mutation in selenium-deficient people? As mentioned earlier, they mutate relatively easily. These particular viruses use RNA rather than DNA as their genetic "master plan." RNA is comparatively less stable. These viruses also lack an ability to correct errors in reproducing themselves.

It is also noteworthy that Dr. Will Taylor of the University of Georgia has discovered that HIV, Ebola and several coxsackie viruses and hemorrhagic fever viruses need to incorporate selenium into structures. He has not yet studied the need of influenza viruses for selenium at this writing. Dr. Taylor suggests that viruses such as Ebola and HIV utilize selenium to produce a selenoprotein which halts HIV replication, thus HIV becomes more invasive in persons who are selenium-deficient.

(As a side note, Dr. J.-C. Hou of Beijing reported at the International Symposium on Human Viral Diseases that only two milligrams of selenium daily for nine days reduced the death rate of viral hemorrhagic fever from 100 percent to 36.6 percent.)

Getting back to the six dramatic days in April 1996, on April 20 at the International Symposium on Human Viral Diseases held in Nonnweiler, Germany, Dr Shu-Yu Yu of Beijing reported that selenium supplementation reduced liver cancer by 35 percent. When selenium supplementation was halted, the liver cancer rate soon returned to normal.

On April 21, also at the Human Viral Disease Symposium, Dr. Luc Montagnier, the discoverer of HIV as the cause of AIDS, and Dr. Marianna Baum of the University of Miami, reported new findings on the relationship between selenium deficiency and decreased immune function in HIV-infected persons. It is a downward spiral as AIDS patients become increasingly selenium deficient and selenium deficiency contributes to further immune function decline. Selenium and the immune system will be discussed later, but there is evidence that selenium-containing proteins are involved in the proper function of T cells, especially CD4 (T4) and CD8 (T8) lymphocytes.

Previously, another clinical trial conducted in China had shown that a daily supplement combining 50 micrograms of selenium, 30 milligrams of vitamin E and 15 milligrams of beta-carotene had reduced the overall death rate by nine percent, the overall cancer death rate by 13 percent, the lung cancer death rate by 45 percent and the stomach cancer death rate by 21 percent.

These dramatic results and robustness of the evidence surprised most nutritionists and medical researchers, but it was frosting on the cake for those of us who have researched selenium for three decades. This book will tell you what you need to know about selenium to live better longer with greatly reduced risks for cancer, heart disease and many other diseases. The dramatic new research will be discussed as well as the background research that provides the foundation for understanding how selenium provides this protection.

WHAT IS SELENIUM AND WHAT DOES IT DO?

Selenium may be an obscure mineral, but it is a trace element (mineral) that is vital to health. Just like the minerals iron and calcium, it is needed in your diet or you will die of malnutrition. If you get at least some, but not enough selenium in your diet, you won't develop outright malnutrition, but will instead have a greater chance of getting cancer, heart disease and arthritis, and of experiencing accelerated aging. Now there is exciting and strong evidence that optimal selenium intake will prevent or reduce the incidence of many deadly diseases. This book explains how proper selenium nutrition can safely improve and lengthen your life.

In 1973, Dr. J. T. Rotruck (then at the University of Wisconsin) and his colleagues discovered that selenium was incorporated into the active region of an antioxidant enzyme, glutathione peroxidase (GPX). The importance of antioxidant enzymes will be discussed, but the point here is that for the first time, there was evidence that selenium was essential for human life. Since then, other selenium-containing enzymes and other important proteins have been discovered. Of particular note are two more antioxidant enzymes, phospholipid hydroperoxide glutathione peroxidase (GPX-PH or GPX-II) and plasma glutathione peroxidase (GPX-P). The fourth selenium-containing enzyme known at this time is a critical enzyme in the thyroid, iodothyronine deiodinase, that helps convert the thyroid hormone T4 (tetraiodothyronine) into T3 (triiodothyronine).

Important selenium-containing proteins that are not enzymes include a selenoprotein in the prostate, prostatic epithelial selenoprotein (PES) and selenoprotein P. PES may help explain why selenium is so critical to preventing prostate cancer. Selenoprotein P is a selenium-containing protein found in blood that is believed to be the major transporter of selenium throughout the body. The selenoprotein P level is very sensitive to the level of selenium in the diet. In addition to transport, it appears that selenoprotein P is a free radical scavenger, as well as being important to mood. Low selenoprotein P levels are associated with depression and schizophrenia.

Another selenoprotein, mitochondrial capsule protein (SeMCP), is important to sperm structure and motility but at this writing its structure has not been fully elucidated. There is evidence that the amino acid selenocysteine should be recognized as the twenty-first amino acid. (Boeck, A., et al., Mol. *Microbiol.*, 1991)

HOW SELENIUM WORKS

Selenium is not a wonder drug. These diseases are all dependent on a common factor called a "free radical," and as just pointed out, selenium is a nutrient that is essential for making a series of antioxidant enzymes called glutathione peroxidases. These antioxidant enzymes are critical to the body's defense against free radicals.

While free radical is not yet a household word, the term is in common use among the most knowledgeable of those interested in improving their lives with optimal nutrition. Our interest in free radicals is due to their high reactivity with vital body components. It is not necessary that we understand what free radicals are or how they do their damage, but it does help if we have a "word picture" conceptual use of the term.

What Are Free Radicals?

Atoms are the smallest particles of a chemical element. These small particles of matter in turn consist of electrons spinning about a nucleus. Those who have had science courses may also remember that these electrons occupy regions of space known as orbitals. Each orbital can hold a maximum of two electrons which are called "paired" electrons.

A free radical is defined simply as any atom, molecule or molecular fragment that exists independently with only one electron in one or more orbitals. Such a lone electron is called an "unpaired" electron. Free radicals are unstable and highly reactive. Free radicals are produced usually as the result of a normal molecule losing or gaining an electron.

Normal body processes (such as metabolism, that utilizes oxygen to turn food into energy) also involve a series of "side reactions" that remove electrons from oxygen atoms one electron at a time, instead of the normal reactions which remove oxygen electrons in pairs. This leaves the oxygen atom with an extra unpaired or lone electron in comparison with a normal atom of oxygen. This is an oxygen-free radical. All that non-chemists have to understand is that a free radical is an active part of a molecule. Basically, a free radical is a chemical species with an odd number of electrons.

We use oxygen to oxidize our food and make energy for us to live. This happens in the mitochondria present in the cytoplasm of most living cells. Several different types of oxygen radicals and other reactive oxygen species can be formed during normal metabolism. Instead of food oxidation in mitochondria happening in the normal two electron steps, about one to two percent of the time, it occurs in a single electron step. The first oxygen radical produced is the superoxide anion radical. This radical can gain an electron and be converted into hydrogen peroxide, which in turn can gain an electron and be converted into a hydroxyl radical. The hydroxyl radical is the most damaging radical to body components.

It is not important for the reader to understand the chemistry of free radicals, unless you are a health professional. If you are, then you may be interested in the discussion for health professionals at the end of this section.

Free Radicals Do Extensive Damage

Since the laws of nature work to restore atoms and molecules to their normal states, this oxygen atom becomes very reactive and seeks to regain an electron by grabbing one from another atom or molecule. A compound or atom having an unpaired (lone) electron is highly reactive and can initiate a chain reaction of thousands of free radical reactions within seconds, unless the free radicals can be deactivated by restoring the electron pairings.

You may not have realized it, but you are familiar with several examples of oxidation and even free radical reactions. When iron rusts, this is oxygen uniting with the iron to form an oxide. When butter or vegetable oil turns rancid, this is lipid peroxidation, a free radical reaction. When paint dries to a hard finish, the new surface is the result of free radical reactions.

We often use the term "reactive oxygen species" (ROS) now to include species other than radicals. As an example, singlet oxygen and hydrogen peroxide have all paired electrons and are not free radicals, but they can damage body components.

Oxygen free radicals, in turn, multiply the damage of the original oxygen molecule. When one free radical captures an electron from another molecule, this in turn creates a new free radical as the second molecule now has a lone, unpaired electron. Of course, this new free radical will seek to capture another electron and become normal again. As you can see, this process could be perpetual. One free radical grabs an electron from another molecule which then becomes a free radical and grabs another electron from still another molecule and on and on. These oxygen free radicals perpetuate many other free radicals setting off a chain of damaging reactions. This damage occurs at the molecular and cellular levels.

There are tens of thousands of free radicals created in the body every second. Critical body components such as DNA, the genetic material that directs the manufacture of each of our cells, can be altered by free radicals. The DNA in each cell may be hit by free radicals 10,000 times a day.

ANTIOXIDANTS

In health, free radical and reactive oxygen species are balanced by our antioxidant defense system, which consists of a few enzymes and several nutrients. Unfortunately, free-radical chain reactions take place in your body countless times a day. To make matters worse, there are other sources of free radicals besides those generated during oxygen metabolism. Cigarette smoke, pollutants, sunlight, radiation, and even emotional stress can cause free radicals and free-radical chain reactions.

Simply put, antioxidants stop free-radical reactions. A more formal definition is that an antioxidant is a substance that, when present at low concentration compared to those of an oxidizable substrate, significantly delays or prevents oxidation of that substrate. The key is that only a small amount is needed to protect a large amount. Antioxidants act catalytically to prevent oxidative harm. A catalyst is a material that speeds up chemical reactions without entering into the reactions.

Our antioxidant defense system consists of antioxidant enzymes, nutrients and compounds that remove metal pro-oxidants from the body. Major antioxidant enzymes include glutathione peroxidase, catalase and superoxide dismutase. Major antioxidant nutrients include vitamins C and E, bioflavonoids, and possibly vitamin A and carotenoids.

The antioxidant enzyme glutathione peroxidase utilizes glutathione to help convert hydrogen peroxide into water, detoxify other peroxides and synthesize prostaglandins.

If a large quantity of oxygen radicals and reactive oxygen species is produced by an acute exposure to oxidants, then the endogenous antioxidant defense mechanisms may be overwhelmed. They will just simply not be able to cope with a load of free radicals, and this will initiate free radicals into crucial biological molecules, small molecules and macromolecules which can be propagated, leading to molecular damage, cross-linking and inactivation of function activity.

We know that selenium is more than an antioxidant, but we have not yet elucidated all of its actions. Later I will discuss selenium's role in enhancing the immune response and in apoptosis (i.e., programmed cell death). Figure 1 illustrates several of the ways in which selenium protects against cancer.

SELENIUM RESEARCH: OVERWHELMING EVIDENCE

I have had the pleasure of researching selenium since 1959. This was long before we knew how it was used by the body and that it was essential to man. All that I knew then, based on the research of Drs. Klaus Schwarz and Milton Scott, who are both now deceased, was that selenium seemed to have an interrelationship with vitamin E in chickens, swine and cattle and that low selenium intakes in these animals appeared to be possibly associated with muscle, heart, kidney and liver necrosis. My book *Selenium as Food and Medicine* (Keats, 1980) was dedicated to Dr.

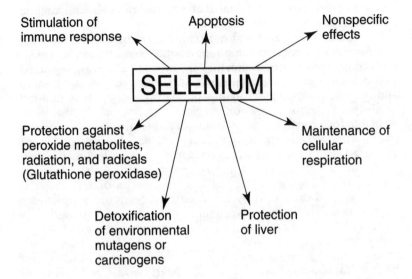

Modified from Schrauzer, G. 1978. *Inorganic and Nutritional Aspects of Cancer*. New York: Plenum Press, p. 330.

Fig. 1. Selenium protects us against cancer via several mechanisms.

Klaus Schwarz and contained an extensive account of his research.

I included selenium in my research because it seemed to me from a theoretical basis that compounds having selenium bonds would form better radiation protectors and free radical scavengers than the sulfur-containing compounds being used in other research.

The concept of antioxidant synergism was new at that time, as were practical amounts of compounds, including nutrients that could extend the lifespan or prevent cancer. Therefore, I patented this concept and formulations in several countries. The original patents have expired long ago, but I and a colleague still have active patents on the use of very effective novel selenium compounds for the prevention of cancer. I will discuss more of my research later in this book.

Ladies' Home Journal (July, 1971), and *Prevention* magazine (December, 1971) published articles about my research which may have been the first times the general public ever read about selenium as a nutrient. Actually, those articles most likely were the first times the public ever read about free radicals and antioxidant nutrients.

About the same time that my 1975 book *Supernutrition: Megavitamin Revolution* informed the public about selenium and its role in protecting against free radicals, the Nutrition 21 company introduced high-selenium yeast in bulk quantities for use by vitamin and mineral supplement companies to produce selenium-rich products. Previously, the only selenium compounds that were available for human use were laboratory specialty products costing over a hundred dollars a gram. Later, in 1984, Nutrition 21 introduced a second low-cost selenium supplement, selenomethionine.

The introduction of low-cost selenium products has made an enormous difference in the health of hundreds of thousands of people around the world. A world without selenium supplements would mean that hundreds of thousands of people would have lower levels of the antioxidant glutathione peroxidase and serious deficiencies in their antioxidant defense mechanisms. They would have impaired immune responses and higher risks of nearly eighty free-radical diseases including cancer, heart disease and premature aging. They would have greater morbidity, increased medical expenses and shorter lives.

Today's researchers are expanding the knowledge produced by the pioneers who preceded them, including Gerald Coombs, Sr., Gerhard Schrauzer, Milton Scott, Klaus Schwarz, Raymond Shamberger, Julian Spallholz, Al Tappel, Orville Levander, Gerald Coombs, Jr., P. Whanger, J. Harr, Clark Griffin and myself. Of these selenium pioneers, only Schrauzer, Shamberger, Spallholz, Whanger, Harr, Griffin and I experimented with selenium as an anticancer nutrient.

Selenium supplements themselves are powerful health protectors, but selenium supplements work even better with other antioxidant partners. My first scientific presentation of my 1960s research showing that selenium in synergistic combination with other nutrient antioxidants (anti-free radical compounds) slowed the aging process in laboratory animals was at the twenty-third Gerontological Society Meeting in Toronto in October 1970. A confirmation of the anti-free radical power of my synergistic formulation was reported in *Pathological Aspects of Cell Membranes* (Academic Press, 1971) by Dr. Al L. Tappel of the University of California at Davis. My first scientific presentation of how selenium and other nutrient antioxidants in synergistic combination prevented cancer appeared in the magazine *American Laboratory* in May 1973. Earlier presentations of my research were in patent applications and news articles. (See the following for a bibliography of reports on my early research on this topic.)

Bibliography of the author's early publications on anti-cancer activity of selenium

IN THE SCIENTIFIC PRESS

1. *The Gerontologist,* 10(3):28, Oct. 1970
2. *Chem. & Eng. News* 48:17 (Oct. 26, 1970). Follow-up: May 10, 1971
3. *Geriatric Focus,* March-April 1971
4. *American Laboratory* 3(4):36–40, April 1971
5. *American Laboratory* 3(5):21–25, May 1971
6. *American Laboratory* 5(6):10–22, May 1973
7. *American Laboratory* 8(4) 32–47, Aug. 1976

U.S. Patent Numbers 39140 and 97011, and patents in various other countries.

IN THE GENERAL INTEREST PRESS

1. *Washington Daily News* p. 5, Sept. 15, 1970
2. *Prevention* 23(12) 104–110, Dec. 1971
3. Reuters News Agency report, Oct. 23, 1970
4. *Ladies' Home Journal* 70, 129–131, July 1971
5. Kugler, H., *Slowing the Aging Process,* Pyramid, NY, 1973
6. Rosenfeld, A., *Prolongevity,* Knopf, NY, 1976

Through the years, we kept hammering away, expanding our experiments, lecturing our colleagues, and daring somebody to prove us wrong. We had to become "activists" to bring this important research before as many scientists as we could. New scientists who entered the field brought with them new skills and capabilities that produced different kinds of evidence that selenium did prevent cancer. Here is a look at the exciting new evidence involving clinical trials, as well as a review of the major epidemiological studies and laboratory studies. You don't have to be a biochemist to understand how strong the evidence is.

THE MECHANISM OF CANCER

"Cancer" is not a single disease, but a group of related diseases. There are more than 100 different types of cancer, but they all are a disease of some of the body's cells. Most cancers involve tumors. The National Cancer Institute explains tumor production as follows:

Healthy cells that make up the body's tissues grow, divide, and replace themselves in an orderly way. This process keeps the body in good repair. Sometimes, however, normal cells lose their ability to limit and direct their growth. They divide too rapidly and grow without any order. Too much tissue is produced and tumors begin to form. Tumors can be either benign or malignant.

Benign tumors are not cancer. They do not spread to other parts of the body and they are seldom a threat to life. Often, benign tumors can be removed by surgery, and they are not likely to return. Malignant tumors are cancer. They can invade and destroy nearby tissue and organs. Cancer cells also can spread (metastasize) to other parts of the body, and form new tumors.

The operative words are "unregulated growth of cells." What causes cells to go astray? Whatever causes cell regulation to go astray is a cause of cancer. Free radicals can damage our genetic material, DNA, and lead to the initiation of cancer. If the DNA is damaged, it can make the wrong stuff or can churn out more than is needed. If the wrong stuff is produced, then the cell becomes altered or mutated. Mutated cells grow fast because they are not confined by quantities of similar cells. If the DNA churns out more cell material than needed, the cell is forced to divide, causing uncontrolled tissue growth.

When a chemical reacts with components of the cell membrane (the "skin" of the cell), the sensors that tell the cell when to divide for growth can be damaged. If a sensor is damaged then cell growth is uncontrolled and that is the *start* of cancer.

Some chemicals inhibit the immune system from functioning properly either directly or indirectly. Although this is not direct damage to the cell's regulatory mechanism, it still involves part of the overall regulation of cell growth. One of the major roles of the immune system is to recognize mutated cells and destroy them. An impaired immune system may let mutated cells develop into cancers.

Cancer development is not a single-step process. Keep in mind that initiating the development process alone does not necessarily lead to cancer. This process only produces a series of independent precancerous cells. The process must be propagated to the point where these precancerous cells reproduce, associate and develop their own blood supply and defense system. If there is no propagation, or if the immune system is activated and destroys these precancerous cells, then no cancer will develop.

The next step in cancer development, called promotion, allows the precancerous cells to reproduce rapidly and change their membrane surface properties to those characteristic of malignant cells. Anything that promotes cell reproduction decreases the chance that repair enzymes will repair (or deactivate) the activated oncogene.

Compounds that increase cell reproduction (also called cell division or mitosis) are commonly called promoters. Cancer researchers call them "epigenetic carcinogens." Many promoters work by damaging growth suppressor genes or by inactivating components of the immune system. Common promoters include alcohol, hormones and polyunsaturated fats.

As an example, smoking and drinking are cocarcinogens in esophageal cancer. Tobacco smoke components initiate the cancer by altering the DNA of the genes and then activating the altered genes. Alcohol becomes a factor because it promotes the development of cancer from precancerous cells.

Even with promotion, proliferating cells will not necessarily develop into cancer. The cell mass must grow large enough to affect body metabolism and start their own blood supply and defense system. This is where the immune system can do its job, unless it is impaired. Sometimes the altered cells are so defective that they just die off.

The answer, then, to the question "How does selenium prevent cancer?" is that it may do so in a variety of ways, including stopping free radical damage, stimulating the immune system, inactivating carcinogens, maintaining cellular respiration, and, as Dr. Larry Clark has suggested, possibly a mechanism involving apoptosis. To achieve full protective effect, it is best to consume adequate amounts of selenium before and during the exposure to carcinogens or excessive free-radical attack, so that selenium can produce adequate glutathione peroxidase to be a shield to prevent the damage that leads to cancer. After exposure to carcinogens, selenium stimulates the immune system and regulates apoptosis.

SELENIUM PREVENTS CANCER

There is no longer any question that selenium prevents cancer! There is even preliminary evidence that selenium may help *cure* some types of cancer. Selenium by itself can't protect all people from all types of cancer, but it definitely reduces the incidence of many types of cancer studied so far in large populations. Some individuals may have genetic weaknesses, or be so undernourished that their immune system can't function, or be so overwhelmed with chemicals that cause cancer that they don't have a chance. However, most people are not so unfortunate, and selenium can reduce their risk of getting cancer by strengthening the immune system and protecting vital cell components against free radical attack.

Nearly six hundred studies have been conducted on selenium and cancer at this writing. Many of those studies are laboratory studies, case-controlled clinical studies and epidemiological studies, which consistently demonstrate that the better the selenium nutriture of a person, the less chance of cancer. In my 1987 book *Selenium Update,* I stated that the evidence for a prophylactic role for selenium is without a doubt stronger than for any other factor, including the nutrient beta-carotene, dietary fiber, crucifers or any other food factor recommended as potential cancer-preventive agents by the American Cancer Society, National Cancer Institute, and other organizations. This is still true today.

The evidence includes the following types of studies:

Clinical: prospective, retrospective
Epidemiological: blood levels, food intake, soils
Ecological: geographical studies of selenium soil content versus disease incidence
Animal: lifespan, spontaneous cancer, carcinogen-induced (dietary, contact), virus-induced (not inoculated, inoculated), transplanted cancer tissue, inoculated cancer cells
Therapeutic: preliminary.

Before the three human clinical trials discussed herein were reported, there were epidemiological (population), ecologic (geographical correlation) and laboratory animal studies to learn from. The protective action of selenium against cancer has been demonstrated in laboratory animals regardless of whether the cancer-causing agents have been introduced in the laboratory or naturally, were chemically or virally induced, inoculated or transplanted.

The clinical confirmation that cancer probability correlates inversely with a person's blood selenium content (the higher the selenium, the less chance of cancer) was shown epidemiologically in the 1970s by Dr. Raymond Shamberger, and clinically in the *New England Journal of Medicine* by Harvard's Dr. W. C. Willett and colleagues in 1983.

There is no need to review the hundreds of studies. Even before the Clark study and the two Chinese studies, there were six especially dramatic and compelling studies that should have caught the attention of all medical researchers and health officials. Here is a brief summary of these nine studies, which are discussed in greater detail in the next section.

1. Clark Clinical Trial: Dr. Larry Clark of the California Cancer Center in Tucson led a team of 33 scientists associated with 23 organizations in a seven-clinic clinical trial involving 1,312 participants. The trial convincingly showed that selenium supplements reduced the incidence of, and mortality from, various cancers. The clinical trial found that 200 micrograms of selenium taken daily in the form of high-selenium yeast reduced the overall cancer death rate by 52 percent and overall cancer incidence by 42 percent. The overall mortality from all causes of death was reduced by 21 percent. Prostate cancer incidence was reduced by 69 percent and colorectal cancer by 64 percent. Lung cancer mortality was significantly reduced 49 percent and lung cancer incidence was reduced by 40 percent, but this latter reduction was not considered to be statistically significant.

2. Yu Chinese Liver Cancer Clinical Trial: As mentioned earlier, Dr. Shu-Yu Yu of Beijing, China found that selenium supplementation reduced liver cancer by 35 percent. When selenium supplementation was halted, the liver cancer rate soon returned to normal.

3. NCI-Chinese Clinical Study: Another clinical study conducted in China showed that supplementation with selenium, vitamin E and beta-carotene reduced total death rate by 9 percent, total cancer death rate by 13 percent, lung cancer death rate by 45 percent, and stomach cancer death rate by 21 percent within five years. The study was not of selenium supplementation alone, but selenium and its symbiotic partner vitamin E probably accounted for most of the protective action.

4. Willett Study: Considering selenium blood levels alone, those persons in *the lowest fifth* of all blood selenium levels *have twice the incidence of cancer* as those in the highest fifth.

5. Yu Chinese Ecologic Study: Total cancer mortality is *three times higher* in persons having *blood selenium values below a certain value* than the incidence of cancer in those above this value.

6. Clark Skin Cancer Study: Considering selenium blood levels alone, those persons in *the lowest tenth* of all blood selenium levels *have six times the incidence of* cancer as those in the highest tenth.

7. Horvath Study: Both selenium and vitamin E are needed together to prevent cancer.

8. Salonen Study: When considering both selenium and vitamin E blood levels, those persons in the lowest third of blood vitamin E level and also having a low blood selenium level had more than

eleven times the incidence of cancer as those in the upper two-thirds of blood vitamin E and selenium levels.

9. Milner Review: Still another researcher concludes that "selenium should be considered not only as a preventive, but also as a therapeutic agent in cancer treatment and may act additively or synergistically with drug and X-ray treatments."

THE MAJOR SELENIUM CANCER STUDIES

Now let's look at more of the details of the nine studies that I feel are representative of the evidence that selenium dramatically and significantly cuts cancer incidence and cancer death.

CLARK CLINICAL STUDY

Even before Dr. Larry Clark's skin cancer study results were published in 1984, he saw the need to conduct a "gold standard" clinical trial to more fully explore the exciting results suggested when comparing blood selenium levels to skin cancer incidence. Starting in 1983, Dr. Clark and colleagues had organized 33 researchers to study patients in seven clinical centers.

A prospective, double-blind, placebo-controlled, randomized clinical trial was designed utilizing volunteers who had previously experienced skin cancer. The trial randomized 1,312 patients having a history of basal cell or squamous cell skin cancer to take either a placebo or 200 microgram selenium supplement daily for an average of five years. In double-blind studies, neither the participants nor the participating researchers know who receives which of the identical-looking pills until after the study is completed. The placebo is merely an inert pill designed to appear the same as the active pill, in this case the selenium supplement. Dr. Clark's clinical trial was the first such clinical trial to properly test whether or not a selenium supplement alone could reduce cancer risk in humans.

The eight researchers who comprised the Safety Monitoring and Review Committee recommended that the clinical study be halted early because they could detect that the selenium-supplemented group had a greatly reduced rate of cancer and felt that it would be unethical to continue to withhold this nutrient from the others receiving the placebo. The blind phase of the study was officially halted with the approval of the National Cancer Institute in January 1996 and the results were tabulated and presented at the 1996 Federation of American Societies for Experimental Biology meeting in Washington D.C. on April 16.

The study is formally called "The Nutritional Prevention of Cancer with Selenium 1983–1993: A Randomized Clinical Trial." The presentation abstract stated:

> Selenium treatment was, however, associated with statistically significant reductions in . . . total and lung cancer mortality, total cancer incidence, colorectal cancer and prostate cancer incidence. Total cancer incidence was 42 percent lower in the selenium group (p <0.001). The consistencies of these associations over time, between study clinics and for the leading cancer sites strongly suggest that, for this cohort of patients, selenium-supplementation had substantial health benefits . . . [These] results support the hypothesis that supplemental selenium can reduce the incidence of, and mortality from, carcinomas of several sites.

Medical statisticians have arbitrarily agreed that a result is called "significant" if there is less than one chance in twenty (5 percent) that the result is due to chance alone and that the result is consistent with the hypothesis being tested. This is expressed mathematically as the parameter, p, having a "p-value" of 0.05 or less (p<0.05). In the above quote from Dr. Clark's abstract, the p-value given was p<0.001. This means that the results are highly significant. There is less than one chance in a thousand that the 42 percent reduction in cancer incidence was due to chance alone. In other words, the cancer reduction was due to the selenium supplements and nothing else.

The oral presentation also revealed that there was a 21 percent reduction in overall mortality, due primarily to the 52 percent reduction in total cancer mortality. The 52 percent reduction in total cancer mortality was highly significant with p<0.0009. However, the 21 percent decrease in overall mortality was considered not quite statistically significant at p=0.07.

Lung cancer mortality, which accounted for about one-half of all cancer deaths, was reduced by a significant 49 percent (p=0.046). Dr. Clark reported that the number of cancer deaths for other types of cancer were insufficient for valid statistical analysis, however, mortality for each site was lower in the selenium-supplemented group.

When considering cancer incidence, as opposed to cancer mortality as we were just discussing, the total cancer incidence was reduced a significant 42 percent (p=0.0003), including a 46 percent lower incidence in total carcinomas (p=0.0001). All seven clinics had lower rates of cancer incidence in the selenium-supplemented group. There was a 69 percent reduction in prostate cancer (p=0.0003), a 64 percent reduction in colorectal cancer (p=0.017), and a 40 percent reduction in lung cancer (p=0.082) incidence. Nine of thirteen categories of cancer showed a lower incidence in the selenium-supplemented group.

The researchers concluded that selenium was having an effect on fairly late stages of cancer. This is because such dramatic reductions were achieved so quickly in such cancers as lung cancer that take decades to develop. The researchers are looking at possible effects of selenium on apoptosis (programmed cell death). If so, this would also have implications for HIV-infected and AIDS patients as will be discussed much later in this book.

No signs of toxicity or adverse effects of selenium were observed in this clinical trial. (Clark, L.C., et al., *FASEB J.*, 10[3] A550 Abst. # 3171, March 8, 1996)

YU CHINESE LIVER CANCER CLINICAL TRIAL

Dr. Shu-Yu Yu of Beijing led a placebo-controlled clinical trial in five townships in the low-selenium soils of Qidong County involving 130,471 persons. An eight-year follow-up showed that those given selenium-supplemented salt had a 35.1 percent reduction in incidence of primary hepatocellular carcinoma (liver cancer). There was no change in liver cancer incidence among those receiving standard salt without added selenium. Upon withdrawal of selenium from the supplemented group, the liver cancer rate began to increase.

A clinical study was then conducted among 226 hepatitis B virus-infected persons provided either 200 micrograms of selenium as high-selenium yeast or a placebo daily for four years. The clinical study found that no liver cancer developed in the 113 persons in the selenium-supplemented group, whereas seven cases of liver cancer developed in the 113 persons in the control group. Again, upon cessation of treatment, liver cancer developed at a rate comparable to that in the control group. Dr. Yu remarked that this demonstrated that a continuous intake of selenium supplements is essential for the prevention of liver cancer.

Hepatitis B infection is strongly associated with liver cancer.

Earlier animal studies by Dr. Yu found that selenium supplementation reduced hepatitis B infection by 77.2 percent and liver precancerous lesions by 75.8 percent (Yu, S.-Y. International Symposium on Human Viral Diseases, Nonnweiler, Germany, April 20, 1996)

NCI-CHINESE CLINICAL STUDY

The U.S. National Cancer Institute sponsored a placebo-controlled, randomized, double-blind clinical study conducted in the Linxian County of China involving 75,000 volunteers. This clinical trial showed that supplementation with 50 micrograms of selenium, 30 International Units of vitamin E and 15 milligrams of beta-carotene reduced total death rate by 9 percent, total cancer death rate by 13 percent, lung cancer death rate by 45 percent, and stomach cancer death rate by 21 percent within five years. All figures have high statistical significance except the lung cancer death rate which involved only 31 deaths due to lung cancer.

The study was not of selenium supplementation alone, but selenium and its symbiotic partner vitamin E probably accounted for most of the protective action. I believe that even better results would have been obtained if 200 micrograms of selenium had been used, and the trial continued for longer than five years. (Blot, W. J., et al.; *JNCI*, 1993 and Blot, W. J., et al.; *NEJM* 331(9):614, 1994).

THE WILLETT STUDY

This study is of major importance not only for its research but for its influence. The Willett study was conducted by a well-respected group of researchers, and the work was done at major centers of learning: Harvard, Johns Hopkins, Duke, the University of Texas and other respected colleges and universities. The results were published in a major medical journal, rather than an obscure scientific periodical. Many physicians read of the importance of blood selenium levels and their relationship to cancer risk for the first time, thanks to this study.

In the Willett study, blood samples had been collected in 1973 from 4480 men from fourteen regions of the United States. At the time of collection of the blood samples, none of the men had detectable signs of cancer. The blood samples were preserved and stored for later analysis.

During the next five years, 111 cases of cancer were detected in the group. The researchers then retrieved the stored blood samples from these men. They also retrieved samples from 210 other men (a control group) who were selected because they matched the newly developed cancer patients in age, sex, race and smoking history. The levels of several nutrients and other factors were compared between the men who developed cancer and those men who remained free of cancer.

One difference stood out as being highly significant. The risk of cancer for subjects in the lowest quintile (fifth) of blood selenium was twice that of subjects in the highest. (W. C. Willett et al., "Prediagnostic Serum Selenium and Risk of Cancer." *Lancet* II:130–4, July 16, 1983)

Dr. Willett and his colleagues attempted to find confirmation in larger populations by using toenail clippings to monitor dietary selenium intake instead of using blood samples. Toenail clippings have relatively poor correlation with dietary selenium intake and the range of dietary intake of the participants was too narrow to produce sufficient variation to notice differences. Therefore these studies could not show the protective effect provided by selenium supplementation.

YU CHINESE ECOLOGIC STUDY

This study, examining the relationship between blood levels of selenium in 1458 healthy adults in 24 regions of China, was led by Dr. Shu-Yu Yu of the Cancer Institute of the Chinese Academy of Medical Sciences in Beijing. The researchers found that there was a statistically significant inverse correlation between age-adjusted cancer death rates and the selenium levels in the blood of local residents. In the areas with high selenium levels, there was significantly lower cancer mortality in both males and females. Total cancer mortality was three times higher in areas where mean blood selenium level was greater than 11 micrograms per deciliter of blood than where it was 8 micrograms per deciliter. (Shu-Yu Yu et al., "Regional Variation of Cancer Mortality Incidence and Its Relation to Selenium Levels in China." *Biological Trace Element Research* 7:21–9, Jan.-Feb. 1985)

THE CLARK SKIN CANCER STUDY

This is the study that convinced Dr. Larry C. Clark to organize the clinical trial discussed earlier. While he was at Cornell, Dr. Clark and his colleagues determined the blood selenium levels in 240 skin cancer patients and compared the results to those from 103 apparently healthy persons living in low-selenium areas. The mean blood selenium level for the skin cancer patients was significantly lower than that of the apparently healthy individuals. After adjusting for age, sun damage to the skin, blood beta-carotene and vitamin A levels, and other factors, the incidence of skin cancer in those persons in the lowest decile (tenth) of blood selenium was 5.8 times as great as those in the highest decile. (L. C. Clark et al., "Plasma Selenium and Skin Neoplasms: A Case Controlled Study." *Nutrition and Cancer* 6:13–21, Jan.-Mar. 1984)

THE HORVATH STUDY

The preceding six studies have dealt with real people, but let's look at an important laboratory animal study for a moment. Most scientific experiments examine one variable at a time to study the effect of just that one variable. This reduces confusion from confounding factors. Yet the body is not a simple laboratory; it is a biologically complex mechanism that functions independently of science's effort to study it. While this booklet is written primarily about selenium, I want to stress that total nutrition is important. If you are grossly deficient in vitamin C or vitamin E, the correct amount of selenium will not make up for all the deficiencies.

However, there is more to the relationship than balanced total nutrition. There is a special synergistic relationship among all the antioxidant nutrients, but especially vitamin E and selenium. This has been the essence of my research and the basis for my patents and publications. In 1983, a sophisticated study was published by Drs. Paula Horvath and Clemet Ip that clarified this synergistic relationship of vitamin E and selenium.

They found that vitamin E and selenium must both be present to prevent the proliferating phase of cancer. Their evidence indicates that it is not the amount of selenium-containing glutathione peroxidase that is critical but the amount of microsomal peroxidase activity, which is stimulated only by the presence of both vitamin E and selenium. (P. M. Horvath and C. Ip, "Synergistic Effect of Vitamin E and Selenium in the Chemoprevention of Mammary Carcinogenesis in Rats." *Cancer Research* 43:5335–41, Nov. 1983)

The message here is that scientists should not be studying the correlation between blood selenium levels alone and cancer but studying the blood levels of vitamin E and selenium together. Studying selenium alone, we do in fact find that there is a substantial reduction in cancer risk with the higher blood levels of selenium. But we find the same relationship with vitamin E and some types of cancer—the more vitamin E in the blood, the lower the incidence of cancer. As an example, consider the report in the *British Journal of Cancer* (49:321–324, 1984) by Dr. N. J. Wald that showed that women in the lowest quintile of vitamin E levels had 5.2 times the risk of cancer as women in the highest. However, when a person's blood is rich in both vitamin E and selenium, the protection given that person is far more than that of adding the vitamin E protection and the selenium protection together.

If a person has a normal blood level of selenium but is very deficient in vitamin E, that person will not have a good defense against cancer. Conversely, if a person is a little low in selenium but well-fortified with vitamin E, that person may be more resistant to cancer.

Since the vitamin E blood level can affect the usefulness of selenium, researchers should be looking at the combined levels, not just a simple selenium level. Once researchers catch on to this relationship, we will see even more dramatic results. This becomes apparent in the next study.

THE FINNISH STUDY

Dr. Jukka Salonen and his colleagues at the University of Kuopio in Finland have been studying over 12,000 Finns for several years. The study is known as the North Karelia Project. Four years after drawing blood samples from these 12,155 persons, 51 had died of cancer. They were matched by age, sex and smoking habits with others and their blood samples were compared.

In this study many factors were examined, but most important, both vitamin E and selenium levels of the blood were examined in combination. The relative risk of cancer mortality for the third of people with blood selenium levels below 47 micrograms per liter of blood compared to those with higher levels was 5.8 to 1. But of more importance is the finding that, for people with low selenium levels who also had vitamin E levels in the lowest of values, risk of death from cancer compared to persons with both selenium and vitamin E levels in the upper two-thirds of values was 11.4 to 1. (U.T. Salonen et al., "Risk of Cancer in Relation to Serum Concentrations of Selenium and Vitamins A and E." *British Medical Journal* 290:417–20, Feb. 9, 1985)

THE MILNER REVIEW

While Dr. John A. Milner was at the University of Illinois he studied selenium and cancer prevention for more than a decade.

Most of Dr. Milner's studies involve transplanting or inoculating cancer cells into mice receiving different levels of selenium. He has found that selenium inhibits the development of such cancer.

Dr. Milner's conclusion is that selenium should be considered not only as a preventive but also as a therapeutic agent in cancer treatment. There is evidence, which will be presented later, that selenium may act additively or synergistically with drug and X-ray treatments. (J. A. Milner, "Selenium and the Transplantable Tumor." *Journal of Agricultural and Food Chemistry* 32:436–42, May-June 1984).

Many of the studies providing the support needed to undertake these studies were based on laboratory animal and human population (epidemiology) studies. These studies will be discussed in later sections, including graphic illustrations of several studies indicating inverse relationship between cancer incidence and selenium intake.

The epidemiological studies by themselves do not prove anything, but combining them with clinical studies and laboratory experiments allows a valid conclusion to be inferred.

NEWS FOR NUTRITIONISTS

These nine studies will eventually have a major impact on cancer research. However, the subject is still new to most researchers and especially so to those who hold fast to the notion that diet and cancer are not related. To begin to educate such "old-line" nutritionists, the conservative nutrition journal *Nutrition Reviews* published an issue devoted to the topic.

Drs. Larry Clark and Gerald Combs, Jr. presented an excellent review of selenium's protective roles against cancer in the November 1985 issue of *Nutrition Reviews*. Their article entitled "Can Dietary Selenium Modify Cancer Risk?" presented convincing evidence (*Nutr. Rev.* 43:325–331, 1985). They provide 97 references for the serious scholar to pursue. Most scientists who bother to read those 97 reference articles will become selenium fans; but there are six times that many in the literature, and they should be read to really understand the depth of the research that supports the hypothesis.

Since *Nutrition Reviews* is widely read by nutritional scientists (keep in mind that nutritionists usually are not regular readers of the journals on cancer research), the importance of selenium may finally be appreciated by nutritionists.

A MEASURE OF SELENIUM

One important aspect of selenium research is that blood or hair levels of selenium are of major importance. It is not just simply a matter of what foods one eats, or how much selenium is present in those foods. Foods vary greatly in their selenium content, and just as important, food selenium varies greatly in its availability to the body. Sometimes the food selenium is present in a form more readily assimilated than other forms, and sometimes the selenium is "tied up" with other compounds such as mercury and is unavailable to the body. This is why selenium supplements are important, as they are bioavailable and in measured doses. The importance of blood levels will become clearer in the following section concerning the use of selenium in cancer therapy.

In the early 1970s, when it was clear that optimal intake of selenium was protective against cancer in laboratory animals, Dr. Gerhard Schrauzer (then at the University of California at San Diego) and I independently called for clinical trials to test for this capability in humans. Dr. Schrauzer had completed a series of experiments that showed that optimal selenium intake could reduce the natural occurrence of breast cancer in mice by nearly 90 percent to only 12 percent of the normal cancer rate (G. Schrauzer and D. Ishmael, *Annals of Clinical Laboratory Science* 4:441–7, 1974). He told the National Cancer Institute as far back as 1978 that the key to cancer prevention lies in assuring adequate selenium intake.

Dr. Schrauzer stated in a 1978 article in *Family Circle* that if every woman in America started taking selenium supplements today or had a high-selenium diet, within a few years the breast cancer rate would decline drastically. He also remarked that if a breast cancer patient has low selenium levels in her blood, her tendency to develop metastases (other tumors) is increased, her possibility for survival is diminished, and her prognosis in general is poorer than if she had normal levels.

These observations were based on the experience of a group led by Dr. K. McConnell and may have been what set several surgeons looking to see if better selenium and other nutrient antioxidant nourishment would improve patients' chances of recovery. And guess what—they do.

There are successful reports in the medical literature going back to the 1950s that were missed even by the selenium pioneers working with laboratory animals. In 1956 four leukemia patients were given an organic compound that contains selenium called selenocystine. In all four cases, a rapid decrease of the total leukocyte count was observed, as well as a reduction in spleen size. The most striking results were obtained in the cases of acute leukemia. (A. Weisberger and L. Suhrland, *Blood* 11:19, 1956)

In 1975, as referred to above, Dr. McConnell noticed a link between the survival times of 110 cancer patients and their blood selenium levels. Dr. McConnell and his colleagues concluded that those patients with the lowest blood selenium levels were more likely to have far-spreading cancer, multiple tumors located in different organ systems and multiple recurrences. (K. McConnell, W. Broghamer, A. Blotcky and O. Hurt, *Journal of Nutrition* 105:1026–31, 1975; W. Broghamer, K. McConnell and A. Blotcky, *Cancer* 37:1384, 1976)

Conversely, those patients whose cancers were confined and who rarely suffered recurrences all had higher (but still subnormal) selenium levels. These same researchers determined in 1978 that the blood selenium levels of breast cancer patients were lower than in women in the same age brackets without breast cancer. (K. McConnell, *Advances in Nutrition Research*, vol. 2, ed. by H. Draper. New York: Plenum Press, 1979, p. 225)

In 1983 I reported in my book *Cancer and Its Nutritional Therapies: Updated Edition* that Dr. Richard Donaldson of the St. Louis Veterans Administration Hospital reported his results to the National Cancer Institute (May 9, 1983) and the Annual Meeting of V.A. Surgeons (1980). These reports showed remarkable results for the use of selenium in treating terminally ill cancer patients and emphasize the need for further large-scale investigation of such use. Dr. Donaldson passed away before he could publish his results in medical journals. Although they must be considered preliminary, the bare facts of this work are important enough to merit mention here.

At the time of Dr. Donaldson's presentation to the National Cancer Institute, he had 140 patients enrolled in his study. According to the data in his letters that I have, all the patients who entered his study were certified as being terminally ill by two physicians after receiving the appropriate conventional therapy for their particular cancer. Some of the patients who entered the program with only weeks to live were alive and well after four years, and with no signs or symptoms of cancer. Not all patients were cured, but all had reduction in tumor size and pain. It is unfortunate that they did not receive the selenium until they were pronounced incurable.

It is important to realize that the dramatic improvements did not occur until sufficient selenium was ingested to bring the patient's blood selenium level up to normal. Sometimes this could be achieved in a few weeks with 200 to 600 micrograms of selenium per day, while other individuals required as much as 2,000 micrograms per day to normalize the blood selenium levels.

Selenium can be toxic in the milligram-per-day range. One milligram is 1,000 times greater than one microgram. The RDA range for selenium is 50 to 200 micrograms per day. However, no signs of toxicity were observed in any patients, even in the autopsies of the 37 patients who were helped, but not cured, by the therapy. Other antioxidant nutrients—vitamins A, C and E—were also used in the program.

Dr. Donaldson administered selenium as a nutritional adjunct to conventional therapy to over 100 patients and reported that those patients generally did much better than expected. As an example, of eight patients with inoperable lung cancer, four were alive and well one year later, with no progression of the disease as shown by X-ray examination. Normally, all eight would have been expected to have died in that time.

Dr. Donaldson gave the patients up to 2000 micrograms of selenium daily, along with other antioxidant nutrients including vitamins A, C. and E. He would adjust the selenium supplementation to produce normal blood selenium levels which often did not occur until 900 to 2000 micrograms per day were given.

It's time to proceed to controlled clinical trials to confirm Dr. Donaldson's research. It may be that few cancer treatments will work in patients with selenium deficiencies. It has been found recently that cancer patients will suddenly respond to conventional therapy—that is, their tumors start to shrink—once their blood levels of selenium have been brought up to normal with supplements.

IGNORED TOO LONG

It is time for even larger clinical trials to consider cancer prevention in terms of selenium. Referring to a conference at the

International Association of Bioinorganic Scientists, the *San Diego Union* reported nearly two decades ago, on January 10, 1979:

> Some of the nation's leading nutritional scientists meeting here said they believe there is sufficient evidence that selenium can reduce some cancers to conduct a national trial of the trace element on humans. . . . A majority of the scientists said they are satisfied that available data is sufficient to indicate that supplementation of the diet with 100 to 200 micrograms of selenium will reduce the occurrence of some cancers in humans. . . . They said in a poll that they favor a human field trial to test the effectiveness of dietary selenium against cancer and suggested the study involve 10,000 to 15,000 persons taking selenium and an equal number taking a placebo.

Early Epidemiology

It was Dr. Raymond Shamberger who discovered in 1969 that the selenium blood levels of cancer patients were low. His laboratory experiments in 1966 had, in fact, indicated the presence of selenium deficiencies in cancer. While the normal level is more than 180 micrograms of selenium per liter of blood, many of the cancer patients he studied had only 120 to 150 micrograms per liter.

Dr. Shamberger, together with Dr. Frost, studied the relationship between selenium in crops and the cancer incidence in people

TABLE 1
Relationship between selenium in crops and breast cancer rate

Selenium crop concentration ppm	Breast cancer rate compared to nutritional average
0.03	11% above average
0.05	9% above average
0.10	20% below average
0.26	20% below average

Source: Shamberger, R. and Frost, D. 1969. *Canadian Medical Association* 100:682.

TABLE 2
Cancer death rates in various areas

Selenium level	(ppm) (soil)	No. states	Cancer death rate (per 100,000)
Very High	above 0.26	6	392
High	0.10 to 0.25	19	430
Medium	0.06 to 0.09	11	450
Low	0.01 to 0.05	20	516

The 1968 55–64 age-specific cancer death rate for the white males in states located in regions of selenium occurrence.

Source: Shamberger, R. May 11–13, 1976. From the *Proceedings of the symposium on selenium-tellurium in the environment.* Univ. of Notre Dame, p. 262.

living in those areas. They found that the lower the level of selenium, the higher the incidence of cancer. The results are clearly revealed in Tables 1 and 2, and in Figures 2 and 3.

In another study, Drs. Shamberger and C. Willis found that healthy persons between the ages of 50 and 71 averaged 217 micrograms of selenium per liter of blood, whereas cancer patients of the same age range averaged only 162 microliters per liter. The worst cancer cases had the lowest selenium levels (137, 139, and 143).

Other epidemiological data reveal that Rapid City, South Dakota, has the lowest cancer rate of any city in the United States. The citizens of that city also report the highest measured blood selenium levels. But in Lima, Ohio, where the cancer rate is twice that of Rapid City, the citizens have only 60 percent of the blood selenium levels disclosed in Rapid City. Note the inverse correlation between selenium level and cancer death rate in Table 3, where soil selenium concentration is compared to cancer incidence rates for the continental United States.

Dr. Schrauzer and his colleagues found that in 27 countries surveyed (see Figure 4) the cancer death rate was inversely proportional to the dietary intake of selenium in the typical diets of those countries. Cancers involved in the study included tumors of the breast, ovary, colon, rectum and prostate, as well as leukemia.

Another study showed that the blood selenium levels of persons from 17 countries were also inversely proportional to the breast cancer rate. For example, in Japan, Thailand, the Philippines and other Eastern countries, where blood selenium levels range from

Figs. 2 and 3. Cancer death rate versus selenium level

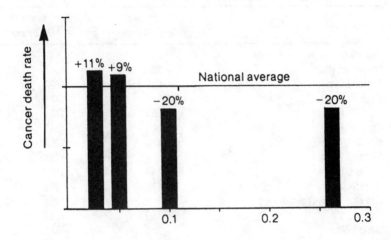

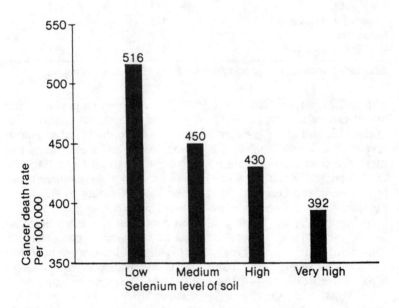

TABLE 3
Selenium concentration in human blood and human cancer death rate in various cities (1962–66)

City	Blood Se (mcg/100 ml)	Cancer deaths per 100,000 pop.
Rapid City, S.D.	25.6	94.0
Cheyenne, Wyo.	23.4	104.0
Spokane, Wash.	23.0	179.0
Fargo, N.D.	21.7	142.0
Little Rock, Ark.	20.1	176.0
Phoenix, Ariz.	19.7	126.7
Meridian, Miss.	19.4	125.0
Missoula, Mont.	19.4	174.0
El Paso, Tex.	19.2	119.0
Jacksonville, Fla.	18.8	199.0
Red Bluff, Calif. (Tehama Co.)	18.2	176.0
Geneva, N.Y.	18.2	172.0
Billings, Mont.	18.0	138.0
Montpelier, Vt. (Wash. Co.)	18.0	164.0
Lubbock, Tex.	17.8	115.0
Lafayette, La.	17.6	145.0
Canandaigua, N.Y. (Ontario Co.)	17.6	168.0
Muncie, Ind.	15.8	169.0
Lima, Ohio	15.7	188.0

Source: Shamberger, R. and Willis, C. June 1971. CRC Reviews.

0.26 to 0.29 parts per million (260 to 290 micrograms per liter), the breast cancer rate was from 0.8 to 8.5 per 100,000 population. But in the U.S. and other Western countries, where the blood selenium levels were lower (0.07 to 0.20 ppm) (70 to 200 micrograms per liter), the breast cancer rate was higher (16.9 to 23.3 per 100,000). Dr. Schrauzer has also analyzed selenium intake versus three types of cancer in the U.S. and has extrapolated this data to suggest a possible "zero cancer incidence." (See Figure 5.)

In Venezuela, the death rate from cancer of the large intestine is 3.06 per 100,000. Venezuela has a high selenium content in its soils, while ours is low. Japan, another high-selenium country, has less breast cancer, as we have noted, and it also enjoys a lower lung cancer death rate—12.64 per 100,000 compared to our rate of per 100,000. This same inverse relationship was found in a 1971 study to correlate with the level of selenium in local cow's milk.

FIG. 4 Relationship of selenium intake and breast cancer mortalities

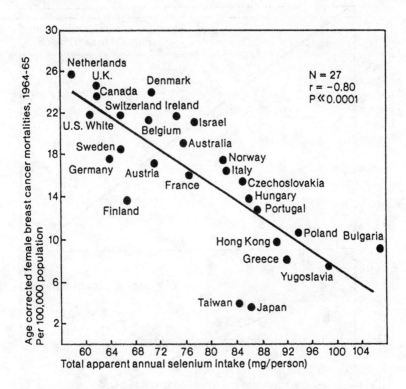

And finally, in a 1972 study made in New Zealand, where selenium was introduced to sheep in anthelmintic drenches, the incidence of sheep cancer practically disappeared.

In 1975, Dr. Birger Jansson of the M. D. Anderson Hospital in Houston presented studies showing the high incidence of colon cancer in low-selenium areas.

It should be pointed out that the foregoing epidemiological data do not necessarily prove cause and effect—they only suggest possible relationships that should be researched further.

FIG. 5 Zero cancer extrapolations

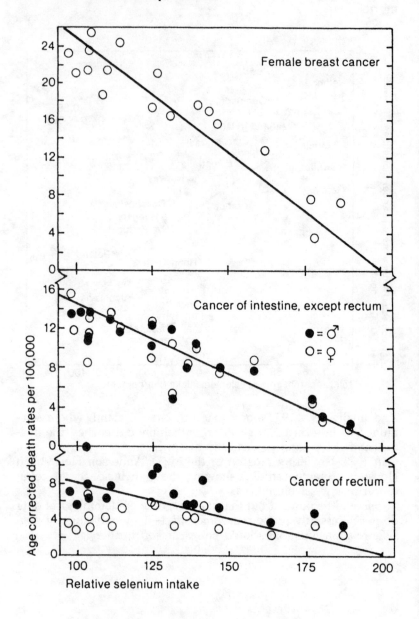

SELENIUM'S INFLUENCE ON DIETARY FATS

Americans tend to eat too many fats. However, fats are not necessarily the total problem. The problem is that we are eating too many fats for the amount of antioxidants such as selenium that we eat. To put it another way, the real problem is the deficiency of selenium and other antioxidants, a situation that is aggravated by the burden of extra dietary fat.

The May 3, 1976, issue of *Chemical and Engineering News* added more information about the protective effect of selenium against dietary fats:

> The association between high selenium levels in the diet and a lower than average cancer rate was suggested in a paper delivered by Dr. Christine S. Wilson, a nutritionist at the University of California, San Francisco. She told the FASEB meeting that high selenium levels in the diet may explain why the breast cancer rate is substantially lower in Asian women than in women from Western countries.

The rate of breast cancer in Asian countries is about one-seventh of the U.S. rate.

After comparing the nutrient content of an average non-Western diet supplying 2500 calories to that of a typical American diet providing the same number of calories, Wilson determined that the Western diets contained about a fourth of the selenium that the Asian diets did. She says that it is also significant that the Asian diets contained much less "easily oxidizable" polyunsaturated fats (7.5 to 8.7 grams a day) than the Western diets did (10 to 30 grams).

The University of California nutritionist hypothesizes that it is the dietary combination of high selenium and low polyunsaturated fatty acids that may be protecting the Asian women against breast cancer. She notes that selenium is a component of the glutathione peroxidase system. Because the enzyme acts to inhibit the oxidation of the unsaturated fats, it blocks the formation of peroxides and free radicals, both of which are believed to trigger various forms of cancers.

The "average" American today ingests about two to three times the amount of polyunsaturated fats than did Americans thirty years

FIG. 6 Selenium is protective against carcinogen-produced cancer

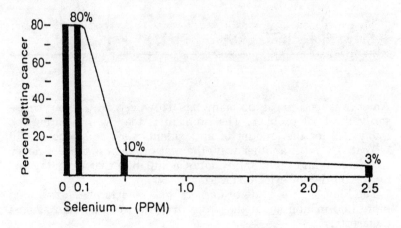

ago. Several studies show proportionate increases in cancer (especially mammary) with increasing dietary polyunsaturate levels.

Pearce and Dayton have reported that humans fed four times more polyunsaturates than were controls developed a significantly higher number of cancers: 31 of 174 as opposed to 17 of 178.

This is not to say that polyunsaturates cause cancer. It is to emphasize the need for a balanced diet and a balance of antioxidants and polyunsaturates. Nor is it to say that polyunsaturates are harmful, or that they should necessarily be reduced if one's diet is unbalanced. It is important to have adequate polyunsaturates for proper nutrition.

But extreme measures to drastically change the polyunsaturated-to-saturated ratio, especially with unbalanced antioxidants, appear to be a great risk, according to available information.

LABORATORY EXPERIMENTS

In a series of experiments in which carcinogens (cancer-causing compounds) were applied to the skin of mice, dietary selenium had a strong protective effect. Selenium deficiency increased the number of mice that developed cancer as well as the number of tumors found in each mouse. Extra dietary selenium reduced both numbers in proportion to the amount of selenium provided.

Comparisons were made by Drs. C. Clayton and L. Baumann between normal diets and those fortified with selenium using azo dyes. Selenium supplementation reduced cancer one-third. Similar tests were run in 1972 by Dr. J. Harr and others with a carcinogen called FAA, and by myself with another carcinogen known as DMBA from 1969 to 1972.

My experiments were made with several antioxidants used in synergistic combination so as to provide animal protection at the lowest total dosage of antioxidants possible. The incidence of stomach cancer to be expected in mice given DMBA is 85 to 90 percent. By adding mixtures of water- and fat-soluble natural and synthetic antioxidants, including selenium, to their diet, that figure was reduced to 5 to 15 percent. In these experiments, the mice I used had been raised from weaning on the experimental diets and they continued to receive the same diets after being fed one dose of the DMBA.

Dr. Harr and his colleagues fed their test animals a diet that included FAA on a constant basis during the entire experiment. Various amounts of selenium were added to the diet concomitantly with the carcinogen to four groups of twenty mice each. Group one received 150 parts per million FAA and 2.5 ppm added selenium; group two received 150 ppm FFA and 0.5 ppm added selenium; group three received 150 ppm FAA and 0.1 ppm added selenium; and group four received 150 ppm FAA and no added selenium. After 210 days, 80 percent of groups three and four had developed cancer, compared to only 10 percent in group two, and 3 percent in group one. The selenium had an undeniably protective effect (see Figure 6).

Drs. Maryce Jacobs and Clark Griffin, while they were working at the University of Nebraska, demonstrated that selenium lowers the incidence of intestinal and rectal tumors in rats exposed to 1.2-dimethylhydrazine and methylazoxy. methanol acetate. They also found that selenium lowers the incidence of liver cancer in rats caused by 3'-methyl-4'-dimethyl-aminoazo-benzene.

In February 1978, at a conference on preventing cancer at the National Cancer Institute in Bethesda, Maryland, considerable emphasis was given to the role of selenium in preventing cancer. Dr. Clark Griffin, now at the M. D. Anderson Hospital and Tumor Institute in Houston, Texas, reported at the time that by adding selenium to drinking water or by feeding selenium in the form of high-selenium yeast, rats that were exposed to three different kinds of cancer-causing chemicals can be protected from colon and liver cancer. Dr. Griffin's group also revealed that selenium can prevent the conversion of potentially cancer-causing chemicals into other harmful forms. Experiments conducted by Dr. Charles R. Shaw, also of M. D. Anderson Hospital, showed that selenium reduced the bowel cancer rate from 87 percent to 40 percent in animals that had been fed carcinogens.

FIG. 7 **Breast cancer reduced by selenium**

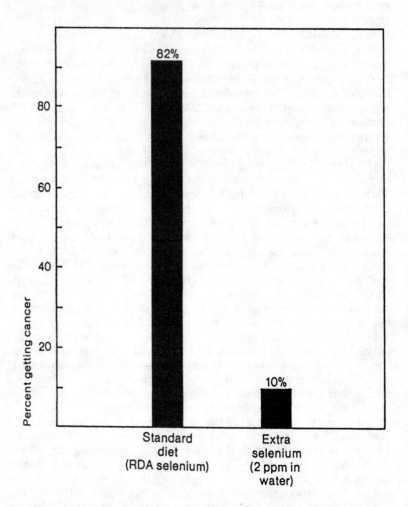

In 1974 Dr. Gerhard Schrauzer tested selenium against the natural cancer incidence in mice. He was able to reduce the incidence of spontaneous breast cancer in susceptible female mice from 82 percent to 10 percent merely by adding traces of selenium to their drinking water. (See Figure 7 for related data.)

In addition to this eightfold reduction of cancer incidence, Dr. Schrauzer's study shows that even among the 10 percent of the selenium-supplemented mice that did develop cancer, the disease appeared 50 percent later than it did among the control animals. Moreover, the tumors were "less malignant, and the survival time of the selenium-supplemented . . . was 50 percent longer. In a less cancer-prone strain of mice, the breast cancer may have been totally prevented."

Dr. P. D. Whanger of Oregon State University studied the effects of selenium on the formation of mammary tumors. He and his colleagues fed groups of twenty to twenty-five mice one of the following diets: lab chow, lab chow plus 0.5 ppm selenium in water, and lab chow plus 2.0 ppm selenium in water. The tumor incidence was 80, 25 and 25 percent respectively (see Figure 8).

FIG. 8 Effect of selenium in preventing breast cancer

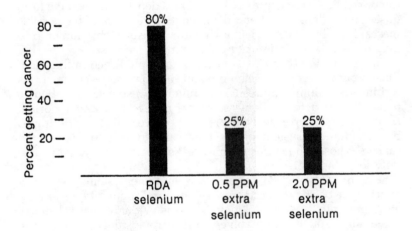

HOW SELENIUM PROTECTS

Selenium is involved in several mechanisms that protect the body from cancer (see Figure 1). It is involved at the first line of defense at the cell membrane as part of the antioxidant enzyme glutathione peroxidase. This protects against free radicals.

One role of selenium is that glutathione peroxidase breaks down the epoxide form from the reaction of carcinogens with aryl hydrocarbon hydroxylase. The compounds that we call carcinogens are only the parent compounds or "procarcinogens." The "active" carcinogens are the epoxides formed within our bodies.

IMMUNE RESPONSE

As mentioned earlier, selenium does more than deactivate free radicals and potential carcinogens. Examining the early research of Dr. John Milner reveals that selenium prevents cancer growth even beyond the initiation stage. Dr. Milner's studies have shown that selenium is capable of retarding the growth of various transplantable tumor cell lines. The degree of inhibition is dependent upon the form of selenium administered and the quantity given. One of the possible mechanisms by which selenium retards transplantable tumor cells may be its action in stimulating the immune response.

There are more than two hundred studies published in the scientific literature regarding selenium and the immune response at this writing. Selenium stimulates the immune response—the body's major line of defense. Some scientists believe that under normal circumstances people develop precancerous cells that are destroyed by the immune system before they can bloom into full-fledged cancers. Those who suffer from a weakened immune system naturally lose this protection.

Selenium has several roles in our immune systems. Selenium affects nonspecific immune indices, humoral immunity, cell-mediated immunity and cytotoxicity. There is evidence that selenium-containing proteins are involved in the proper function of T cells, especially CD4 (T4) and CD8 (T8) lymphocytes. Dr. Will Taylor of the University of Georgia has found that selenoproteins are encoded in the +1 reading frame overlapping the human CD4, CD8, and HLA-DR genes. Cytokines such as the Tumor Necrosis Factor (TNF) are related to selenium levels. Nuclear Factor kappa-B is also affected by selenium and other antioxidant deficiencies.

APOPTOSIS

Apoptosis is the programmed death of cells. There are a dozen research reports in the scientific literature at this writing suggesting that selenium levels are a factor in apoptosis. Oxidative stress in general appears to contribute to apoptosis, but in 1994, researchers in Glasgow postulated that the initial metabolite of selenium, seleno-glutathione (SDG), as well as other organic selenium compounds formed in the body such as dimethyl selenoxide (DMS) and seleno-methyl selenocysteine (SMS), are powerful inhibitors of uncontrolled cell growth. (Lanfear, J., et al., Carcinogenesis 15(7)1387–92, 1994). The researchers found that SDG induces apoptosis and might explain the chemopreventive effects of selenium.

SELENIUM AND AIDS

Apoptosis is also of interest in HIV infection and AIDS. HIV-infected persons become deficient in the selenium-containing anti-oxidant enzyme glutathione peroxidase, which leads to increased levels of lipid hydroperoxides which, in turn, induce apoptosis.

Dr. Will Taylor of the University of Georgia postulates, the more selenium present, the less replication and the more confined the HIV, and the less selenium present, the more replication and viru-lence of the HIV.

When HIV infects a cell, it takes over and manipulates the host cell for the benefit of the HIV. It is as if HIV uses selenium as an indicator of when HIV should replicate, or when it is opportune to replicate, and invade other cells. HIV incorporates selenium that is present in the cell into a selenoprotein that halts HIV replication. This selenoprotein is analogous to a "birth control pill" for the virus. When the selenium content of a HIV-infected cell falls below the level required to make enough of this selenoprotein, then the HIV replicates proportionally.

When host-selenium levels fall, the cell itself becomes vulnerable to apoptosis (programmed cell death) at the same time that the immune system's ability to destroy HIV is diminished. HIV then replicates in order to avoid being destroyed itself when the cell dies, and invades neighboring cells freely without a threat of being destroyed by the weakend immune system.

In 1985, Dr. Brad Dworkin and colleagues at the New York Medi-cal College in Valhalla reported that AIDS patients had significantly diminished plasma selenium levels, and that these levels influenced their immune function. (Dworkin, B., et al., Amer. J. Gastro. 774,

1985). In 1989, Dr. Dworkin and colleagues, as well as French researchers led by Dr. Jean-Fabien Zazzo, determined that the AIDS cardiomyopathy was due to selenium deficiency.

In 1991, researchers in Rome led by Dr. Augusto Cirelli reported that selenium supplementation decreased symptoms in AIDS patients. Also in 1991, researchers at the University of Miami School of Medicine led by Dr. Marianna Baum reported that in spite of adequate dietary intake, more than half of the HIV-infected males that they studied had marginally low plasma selenium levels. Their findings also established that low plasma selenium levels are present during the early stages of HIV infection. (Mantero-Atienza, E., et al., *J. Parenteral Enteral Nutr.* 15(6) 693, 1991)

In 1994, Dr. B. Dworkin reported that selenium deficiency was associated with myopathy, cardiomyopathy and immune dysfunction including oral candidiasis, impaired phagocytic function and decreased CD4 T cells. He reported that in HIV-infected persons, both plasma selenium and glutathione peroxidase levels significantly correlated with total lymphocyte counts. (Dworkin, B, *Chemico-Biol. Interactions* 91:181–6, 1994)

At the start of this book, I mentioned the research of Dr. Marianna Baum of the University of Miami. Dr. Baum's group studied 216 HIV-infected persons from different risk groups and found significant relationships between the level of selenium and immune function in all groups. Dr. Baum told the International Symposium on Human Viral Diseases in Nonnweiler, Germany that "selenium supplementation may offer the potential to moderate immunosuppression and slow HIV-1 disease progression." The group had begun a preliminary pilot study and had already determined that selenium supplementation had produced symptomatic improvements, but as yet, no changes in immunological parameters. The study is continuing.

SELENIUM SUPPLEMENTS

Many health professionals still do not comprehend that there are only two ways to ensure that a diet contains an adequate level of selenium; either directly analyze the foods for selenium content and keep adding foods until the required amount of selenium is

FIG. 9 **Geographical distribution of selenium in the United States**

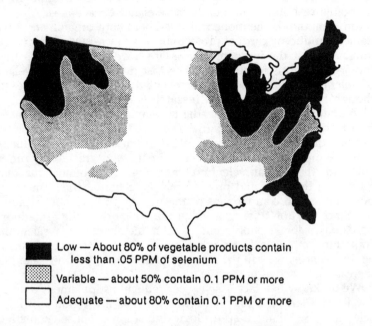

Low — About 80% of vegetable products contain less than .05 PPM of selenium

Variable — about 50% contain 0.1 PPM or more

Adequate — about 80% contain 0.1 PPM or more

eaten, or, rely on selenium supplementation. Choosing a good and varied diet, rich in fruits and vegetables, won't do it. If the soil in which the fruits and vegetables are grown is selenium-poor, then the fruits and vegetables will be low in or devoid of selenium. Fruits and vegetables don't need selenium to grow. Man and other animals need selenium—not plants. If the meat is from livestock fed feed from low-selenium soil, then the meat is low in or devoid of selenium. Figure 9 illustrates the fact that there are vast areas of selenium-poor soils in the U.S.

Health professionals understand this relationship with the mineral iodine, yet many are unaware of the relationship of soil selenium content with food selenium content. Health professionals insist that our table salt be iodized—even in normal iodine-containing soils. Yet health professionals want to believe that a good diet has all the nutrients, including selenium, that we need. This is so if you interpret "need" to mean need for poor health. It is not so if you interpret "need" to mean optimal health, free of cancer.

I have used various forms of selenium in my laboratory animal studies from 1960 through 1989. I found that of the selenium-containing compounds that I tested, high-selenium yeast and the selenium-containing amino acids, selenomethionine and selenocys-teine, as well as the methylated selenides, were far superior to inorganic selenium compounds of selenates or selenites. All sele-nium compounds mentioned in this book, both organic and inor-ganic, produced cancer protection and increased health and longevity, but the inorganic forms are toxic at lower levels than the organic forms mentioned. Since that time, amino acid chelates of selenium have become available and should be effective, al-though I have not conducted research with them.

Vitamin C tends to convert the inorganic selenites and selenates into an ineffective metal form. The scientific literature supports the fact that the organic forms of selenium are more bioavailable than the organic forms, but since the pathways are slightly differ-ent and the actual selenium-containing compounds that are formed in the body and provide the protection against cancer are not established, comparisons of the metabolism are not meaning-ful. Since we don't know whether one or more of the glutathione peroxidases, or selenoprotein P, or a metabolite of a selenium-containing amino acid or whatever is the true protective agent against cancer, we can't measure which dietary form of selenium increases the active protective agent.

We do know one fact—the cancer-reducing supplement used by Dr. Larry Clark and his colleagues was 200 micrograms of high-selenium yeast daily. It seems logical to assume that selenomethio-nine would also offer cancer protection, but this has not been established yet.

The form of selenium-containing amino acids that the body can best use are the "L-" forms, such as L-selenomethionine. The "D-" forms are degraded to inorganic selenium and are only about one-fifth as bioavailable as the "L-" forms.

Studies in New Zealand determined that L-selenomethionine was at least 75 percent bioavailable, compared to 59 percent for sodium selenite. Blood selenium rose more quickly and didn't pla-teau as early with selenomethionine than with sodium selenite.

Dr. P. Whanger of Oregon State University has spent many years studying the effectiveness of several forms of selenium sup-plements. He has published several papers on this subject through the years utilizing various laboratory animals and human clinical trials. His findings are that "the selenium concentrations in all blood fractions increased at a faster rate (two- to three-fold) in those taking selenomethionine than in those taking selenate." (But-ler, J. A., Whanger, P., et al., *Amer. J. Clin. Nutr.* 53:748–54, 1991)

Any selenium-containing amino acids that are incorporated into body proteins are later degraded after normal body protein turnover.

The selenium is then released back into the selenium pool where it may be incorporated into new selenium proteins or excreted Thus, selenium-containing amino acids produce a "time-release" effect—some of the selenium is immediately used to form selenium-containing body compounds while some of the selenium is temporarily incorporated into structural proteins and made available again later.

In a Finnish study, again selenomethionine raised blood levels higher and remained in the blood longer than inorganic selenium. In 1984, a MIT study determined that organic forms of selenium are able to increase the body pool size about 70 percent more effectively than inorganic selenite.

DOSAGE

The amount of selenium in the diet varies with the food selected and the amount of selenium in the soil that particular food was grown in (or the amount of selenium in the feed of the animal). Food tables are of no value for estimating the selenium content of a diet as individual servings of food vary by as much as a thousandfold.

The RDA range established was 50 to 200 micrograms of selenium daily. The latest RDA at this writing is 55 micrograms for women and 70 micrograms daily for men. Some populations average about 600 micrograms and more per day, but the U.S. average is reported to be about 100 micrograms daily—but it depends on where the food was grown.

People eating good diets and living in or near selenium-rich soils should still consider taking another 50 micrograms of selenium daily. Those eating average diets and living in selenium-poor regions should consider taking 100 micrograms of selenium daily as a supplement. Those who eat a poor diet or are at risk of cancer should consider taking 200 micrograms of selenium daily as a supplement. Cancer patients should take whatever amount of selenium is necessary to restore their blood levels to normal.

SAFETY

All things can be toxic at a sufficiently high dosage. Even water or oxygen can have toxic effects at very high levels for long periods of time. It is the dose that makes the poison. Selenium is a mineral that is *essential* in small amounts and it can be toxic at amounts that are small in comparison to vitamins or many other minerals.

Organic forms of selenium are toxic at levels in the vicinity of 3,500 micrograms (3.5 milligrams) daily. Inorganic forms of selenium may be toxic at one-third that level. There is no reason to approach these toxic amounts. There may be even more subtle adverse effects at lower levels than the toxic amounts mentioned above.

In the previous section I mentioned that many Japanese average 600 micrograms of selenium daily. In Northern Greenland, many residents consume about 1,300 micrograms of selenium daily. In China, some residents took 1,100 micrograms of selenium daily when they learned that it protected them from certain selenium-deficiency diseases (including Keshan disease) endemic to their area. They developed thickened fingernails and a garlic-like breath.

There is a report of one case in which a women consumed 2,400,000 micrograms of selenium daily for seventy-five days with only mild and reversible adverse effects. This is 12,000 times the upper range of the original RDA.

Dr. Donaldson gave his cancer patients high amounts of selenium, but he monitored the blood level of selenium to ensure that toxic levels weren't reached. Dr. Hou gave persons with hemorrhagic fever two milligrams (two thousand micrograms) of selenium daily to save them from dying. But he only had to use this much selenium for nine days.

The amounts needed for good health and protection against cancer are in or very near the original RDA range, which is far from being toxic. Selenium is essential to life and the diet cannot be relied upon to provide optimal levels of this vital nutrient. Sensible levels of selenium supplementation are the only assurance that you achieve optimal selenium nutriture.